Georjane de Melo Castro Gondim
Linda Brenna R. Araújo
Thereza R. B. Cavalcante

Water stress tolerance in local cowpea varieties

Georjane de Melo Castro Gondim
Linda Brenna R. Araújo
Thereza R. B. Cavalcante

Water stress tolerance in local cowpea varieties

Study carried out on seedlings

ScienciaScripts

Imprint

Any brand names and product names mentioned in this book are subject to trademark, brand or patent protection and are trademarks or registered trademarks of their respective holders. The use of brand names, product names, common names, trade names, product descriptions etc. even without a particular marking in this work is in no way to be construed to mean that such names may be regarded as unrestricted in respect of trademark and brand protection legislation and could thus be used by anyone.

Cover image: www.ingimage.com

This book is a translation from the original published under ISBN 978-3-330-77388-2.

Publisher:
Sciencia Scripts
is a trademark of
Dodo Books Indian Ocean Ltd. and OmniScriptum S.R.L publishing group

120 High Road, East Finchley, London, N2 9ED, United Kingdom
Str. Armeneasca 28/1, office 1, Chisinau MD-2012, Republic of Moldova, Europe
Managing Directors: Ieva Konstantinova, Victoria Ursu
info@omniscriptum.com

Printed at: see last page
ISBN: 978-620-3-25857-8

SUMMARY

1 INTRODUCTION

The cowpea (*Vigna unguiculata* (L.) Walp) is a dicotyledonous plant belonging to the Fabaceae family and has many popular names. In the northeast of Brazil, it is also known as string bean and macassar bean, while in the north it is popularly called road bean, beach bean and colony bean (FREIRE FILHO; CARDOSO; ARAÙJO, 1983). In these regions, which are the largest producers of the legume (CONAB, 2016), its cultivation is mainly carried out by family farmers, who still use traditional production practices (FREIREFILHO, 2011).

According to Coelho *et al.* (2014), cowpeas are considered to be one of the main sources of vegetable protein. The crop has an average of 23 to 25% protein, all the essential amino acids in its composition, around 62% carbohydrates, vitamins, minerals, dietary fiber, a low amount of hard fat, with an average oil content of 2%, and it also does not contain cholesterol (EM-BRAPA, 2002).

Due to its high nutritional value, this crop is widely cultivated for the production of dried or green kernels and for human consumption, *in fresh*, canned or dehydrated form, as well as other forms of consumption, such as acarajé (ALVES *et al.*, 2009), and its leaves and branches can also be used for animal feed in the production of hay, silage, flour and green fodder (SILVA & OLIVEIRA, 1992). In addition, due to its ability to grow in poorly fertile soils, as well as its hardiness (OLIVEIRA *et al.*, 2002), the cowpea can be used as a green fertilizer and soil protector (EMBRAPA, 2002), incorporated into the soil, providing organic matter and aiding in its recovery, especially in relation to naturally poor soils or soils depleted by intensive use, which are very common in the northeast of Brazil (OLIVEIRA & CARVALHO, 1988, *apud* OLIVEIRA *etal.*, 2002).

In the semi-arid northeastern region, it is one of the main subsistence crops, a basic component in the diet of the population of this region and an important source of alternative income (OLIVEIRA *et al.*, 2002), especially for the part of the population with the lowest purchasing power and for small farmers, and is therefore not only an important component of the diet, but also an important socio-economic component.

However, this region is especially characterized by the occurrence of high temperatures (above 20 ^{0}C of the annual averages), scarce rainfall (between 200 and 800 mm) and water deficit (ARAÙJO, 2011), characteristics that favor the occurrence of drought conditions.

Among the abiotic stresses that affect plant development, such as nutritional and temperature stress, water stress has great significance for plant development. Water deficiency is a common problem in the production of many crops and can have a strong negative impact on both growth and development (LECOEUR & SINCLAIR 1996 *apud* SANTOS & CARLESSO, 1998) and, consequently, generate losses in grain production.

Tolerance to water stress is an important plant defense to maintain production under conditions of low water availability, especially in areas such as the semi-arid region, where rainfall distribution is very irregular and summers are long. It is therefore recommended to use hardier varieties that are tolerant to water stress and have a greater ability to recover after periods of drought (BASTOS *et al.*, 2011).

Based on these factors and considering that the cowpea crop is predominantly grown in a rainfed system, it is important to evaluate the tolerance of this crop to water stress, especially its local varieties, which are largely grown by family farmers in the northeast of Brazil.

With this in mind, the aim of this work was to evaluate different local varieties of cowpea for their tolerance to water stress.

2 LITERATURE REVIEW

2.1 Origin and expansion

The cowpea is of African origin and was introduced to Brazil by Portuguese settlers in the second half of the 16th century, first in the state of Bahia, dispersing to the rest of the country (FREIRE FILHO *et al.*, 2011a). In Acre, for example, it was introduced in the mid-18th century by immigrants from the Northeast who were in the Amazon region (EMBRAPA, 1987, *apud* EMBRAPA, 1997).

Produced in Africa, South America, Asia and Brazil, the legume is widely grown in the northeast and north (SEAB, 2015). Its cultivation is booming in the country and, in recent years, it has also found a market in the central-western region (COSTA, 2010), especially in the state of Mato Grosso (FREIRE FILHO *et al.*, 2011a), where it has been gaining ground, "due to the development of cultivars with characteristics that favor mechanized cultivation" (ALMEIDA, 2014).

Figure 1 - Distribution of Brazilian cowpea production.

Source: Freire Filho *et al.* (2011a).

2.2 Forms of cultivation

The cowpea can be grown mainly in two ways, alone or in consortium with other crops, such as coffee, corn and sorghum (COELHO *et al.*, 2014), in two planting seasons. The first crop is planted at the beginning of the rainy season, from November to March, and accounts for around 71% of the average annual production, while the second occurs from April to August, at the end of the rainy season, accounting for 29% (AGEITEC, 2016).

In Brazil's producing states, it is mainly grown on small areas and in rainfed conditions, in consortium with corn or cassava (ALVES *et al.*, 2009), or with both at the same time, associated with perennial crops, fallow crops and also irrigated (AGEITEC, 2016).

The main advantages of intercropping compared to monocropping are the promotion of greater productive stability and greater use of natural resources, better use of land and labor, better exploitation of water and nutrients, greater efficiency in weed control, reduction of soil erosion and the production of more than one food (AZEVEDO *et al.*, 1997; MATTOS, 1993 *apud* ALVES *et al.*, 2009).

2.3 Production and consumption in Brazil

The bean crop (*Phaseolus vulgaris*) is produced practically all over the country, with around 85% of its production concentrated in the states of Paraná, Minas Gerais, Bahia, São Paulo, Goias, Santa Catarina, Rio Grande do Sul, Ceara, Pernambuco and Para, responsible for around 3.0 million tons/3 harvests (UNIFEIJÂO, 2016). According to data from Unifeijâo (2016), despite the fact that a large part of the production activity is carried out by small rural producers and the extensive use of low-tech labor, the northeast region is responsible for 30% of the world's bean production.

The legume genera produced in the country are Phaseolus (carioca and black), which is more widely grown in the south-central region, and Vigna (cowpea), which is more widely produced in the north and northeast, where it is more popular. Bahia is the largest producer in the northeast, with a 38% share of production, followed by the state of Ceara (19%) - where the municipalities of Crato, Juazeiro do Norte and Jaguaribe stand out - Paraiba (13%) and Pernambuco (8%) (UNIFEIJÂO, 2016). In the North, the states of Amapa, Para, Rondonia and Roraima are the main producers (AGEITEC, 2016).

In the last two harvests, the northeast region was the largest producer of cowpeas, with production of 152,300 tons (t) in the first, with Piaui (87,600 t), Bahia (53,600 t) and Maranhao (11,100 t) as producing states (Table 1), and 189.000 t in the second, with the state of Ceara being the largest producer, responsible for a production of 107,900 tons, followed by the states of Maranhao (25,100 t), Pernambuco (21,100 t), Paraiba (19,100 t), Rio Grande do Norte (13,800 t) and Piaui (2,700 t) (Table 2).

Table 1 - Comparison between area, productivity and production of cowpeas in the first harvest.

REGION/UF	Area (thousand ha)	Yield (kg/ha)	Production (thousand t)
	15/16 harvest (A)	Crop year 15/16(B)	Crop year 15/16(C)
NORTH	**3,7**	**594**	**2,2**
TO	3,7	594	2,2
NORTHEAST	377,0	**404**	152,3
MA	23.8	468	**11,1**
PL	211,7	**414**	37,6
BA	141,5	379	53,6
MIDWEST	1,6	**1,133**	1,7
LIT	1.5	1.133	1,7
SOUTHEAST	0,6	900	0,5
MG	**0,6**	900	**0,5**
NORTH/NORTHEAST	300,7	**406**	154,5
CENTER-SOUTH	**2,1**	1.066	**2,2**
BRAZIL	**382,8**	**410**	**156,7**

Source: CONAB (2016).

Table 2 - Comparison between area, productivity and production of cowpeas in the second harvest.

REGION/UF	Area (thousand ha) 15/16 harvest (A)	Yield (kg/ha) Crop year 15/16(B)	Production (thousand t) Crop year 15/16(C)
NORTH	**1,1**	**727**	**0.8**
RO	1,1	727	0,8
NORTHEAST	**884.0**	**285**	**189.0**
MA	45,7	**549**	25,1
PI	3,0	**900**	2,7
EC	395,9	271	1D7,2
RN	36,2	380	13,8
PB	64,8	294	13,1
PE	118,4	178	**21,1**
MIDWEST	**149,4**	**1.260**	**188.3**
MT	148,0	1.258	186,2
GO	1,4	1.500	**2,1**
NORTH/NORTHEAST	665.1	**285**	**189.8**
CENTER-SOUTH	**149,4**	**1.280**	188.3
BRAZIL	**814,5**	**484**	**378.1**

Source: CONAB (2016).

Despite being among the world's largest producers of cowpeas (WANDER, 2013), the average productivity of the crop in Brazil is considered low, at around 366 kg ha^{-1}, which can be attributed mainly to the low level of technology used in production, given that in states such as Amazonas, Goias, Mato Grosso and Mato Grosso do Sul, where technologies are used that enable the crop's productive potential to be expressed, there are yields of over 1,000 kg ha$^{(-1)}$) (ALMEIDA, 2014).000 kg ha^{-1} (ALMEIDA, 2014).

The cowpea production chain is simple. After its production, the crop can be used to produce inputs for domestic or foreign consumption, for the producer's own consumption, and its grains can also be sent to the agro-industry, where they will be processed and then reach the final consumer (FIGURE 2).

Figure 2 - Schematization of the cowpea production chain.

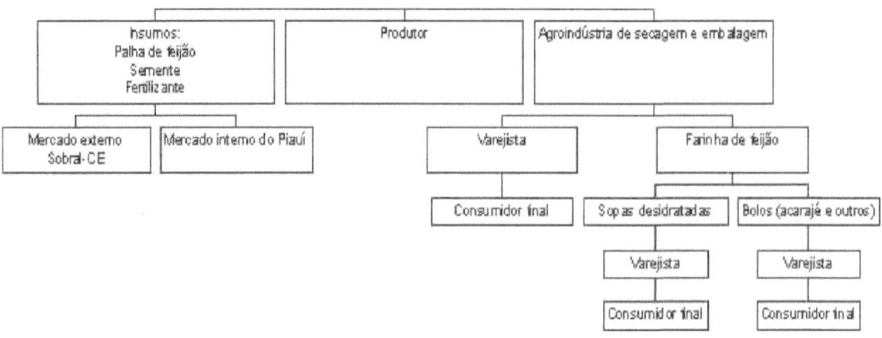

Source: AGEITEC (2016).

Almost half of northeastern production is destined for the producer's own family consumption and, when sold, there is little industrialization in the post-harvest process, but it has been observed that in recent decades the profile of Brazilian rural producers has been changing, especially with regard to mechanization in the field, especially during the harvest phase (AGEITEC, 2016).

Currently, cowpeas are produced in fully mechanized fields and are increasingly reaching the major commercial and consumer centers in other Brazilian regions, such as the Midwest and Southeast, as well as abroad (EMBRAPA, 2011).

The expansion of production areas, the emergence of a new producer profile and consumer market have led to the emergence of new demands for cowpeas, which has broadened the objectives of genetic improvement of this crop (EMBRAPA, 2011).

2.4 Breeding

2.4.1 *A brief history*

In a way, it can be said that the genetic improvement of cowpeas in Brazil began in the second half of the 16th century, when the first cultivars, now known as crioulas or locais, were introduced into the country, although they were exotic, constituting the basic germplasm of Brazil, so that the first genetically improved cultivars emerged from them (FREIRE FILHO *et al.*, 2011b).

Scientific research into this crop only began in 1903, when the book "Os feijôes de macassar", by Gustavo D'Utra, was published in the city of Sao Paulo, which dealt with relevant topics such as its use as a green manure, nutritional value for humans and animals, cultivars and chemical composition (FREIRE FILHO *et al.*, 2011b).

It was only in 1925, however, that the first scientific paper on crop improvement was published, in which the author, Henrique Lôbbe, evaluated 12 cultivars of cowpea. Thus, it can be inferred that genetic improvement probably began in Brazil in 1925 (LÔBBE, 1925 *apud* FREIRE FILHO *et al.*, 2011b).

In summary, genetic improvement in Brazil can be divided into four distinct phases: the first phase lasted 38 years, beginning in 1925 with the publication of Lôbbe's work and ending in 1963, with Ponte's work standing out, in which results were obtained from experiments with cowpeas carried out at the Instituto Agronòmico do Norte (IAN), in the state of Para, in 1962. During this phase, the work was carried out in isolation and had no continuity or correlation with other work. As a result, several introductions were made, but there was no information about the indication of cultivars (FREIRE FILHO *et al.*, 2011b).

The second phase covered a period of 10 years, beginning in 1963, when the Brazilian Bean Commission (CBF) was set up within the National Department of Agricultural Research and Experimentation (DNPEA), and ending in 1973. After the creation of the CBF, research actions related to the genetic improvement of cowpea began to take place in a more participatory and articulated manner, through the DNPEA's regional institutes.

In 1963, the collection, introduction and evaluation of cowpea germplasm began and, after that, the Federal University of Ceara (UFC) began trials of competing cultivars (FREIRE FILHO *et al.*, 2011b).

In 1968, Krutmam *et al.* (1968 *apud* FREIRE FILHO *et al.*, 2011b) published a paper , from which local cowpea cultivars were characterized and introduced. During this period, the first improved cowpea cultivars were released. In the same year, the Seridó cultivar was launched in the state of Ceara, which was obtained through mass selection on the local cultivar of the same name (FREIRE FILHO *et al.*, 2011b).

In 1969, the IPEAN-V-69 cultivar was launched in the state of Para, obtained from the local cultivar 40 Dias, through selection of individual plants with progeny testing (PONTE; LIBONATI, 1969 *apud* FREIRE FILHO *et al.*, 2011b). Also in 1969, the Western Amazon Agricultural Research and Experimentation Institute (IPEAAOC) was founded. During this phase, research into cowpeas began to be carried out at regional level and the integration of regional research institutes with universities was strengthened, in addition to integration with state research institutes, regional credit and rural assistance associations, the National Department for Works Against Droughts (DNOCS) and state agriculture secretariats (FREIRE FILHO *et al.*, 2011b). In addition, the institutes disseminated their work to the experimental stations of the Ministry of Agriculture, located in the states of their respective regions.

The following year, Paiva *et al.* (1970 *apud* FREIRE FILHO *et al.*, 2011b) presented the results of their work on experimentation, plant health and improvement with beans (*Vigna sinensis*) in the state of Ceara. Also in 1970, in Piaui, the results of evaluations of competing cultivars were published (RIBEIRO; MELO, 1970 *apud* FREIRE FILHO *et al.*, 2011b).

The third phase began with the creation of the Brazilian Agricultural Research Corporation (Embrapa) in 1973, and lasted until 1991. In 1974, the National Rice and Bean Research Center (CNPAF) was founded, and in 1977 an agreement was made between Embrapa and the *International Institute of Tropical Agriculture* (IITA), from which an exclusive research team for cowpeas was set up and linked to a national research network for the crop (GUAZZELLI, 1988 *apud* FREIRE FILHO *et al.*, 2011b). Soon after, the National Bean Research Program was organized by CNPAF (EMBRAPA, 1981 *apud* FREIRE FILHO *et al.*, 2011b). During this phase, the national cowpea research network was consolidated, which included Embrapa's decentralized units, state companies, state research institutes and universities, and the components of the National Agricultural Research System (SNPA).

The national breeding research network functioned as follows: work on selecting parents, crosses, generation advances and preliminary trials were carried out at the CNPAF. Preliminary trials were also carried out at the Teresina Research Execution Unit (UEPAE Teresina), the Ceara Agricultural Research Company (EPACE) and the Pernambuco Agricultural Research Company (IPA). The advanced trials were carried out by all the institutions in the research network.

According to Watt *et al.* (1979 *apud* FREIRE FILHO *et al.*, 2011b), during the period from 1977 to 1983, the breeding method used was the genealogical method, while in the period from 1984 to 1981, Kueneman and Guazzelli (1987 *apud* FREIRE FILHO *et al.*, 2011b) state that the single pod progeny method (FEHR *et al.*, 1987) was used. During this phase, the National Bean-Caup Research Meetings (RENACs) began.

The fourth and final phase began 25 years ago, in 1991, when coordination of the National Cowpea Program passed to Embrapa-Meio Norte, and lasts until the present day. The objectives of genetic improvement during this phase focused especially on improving plant architecture, grain quality and adaptation to cerrado conditions (FREIRE FILHO *et al.*, 2001). In 2006, the First National Congress on Cowpea (I CONAC) took place.

At present, research into cowpeas is mainly carried out in the north, northeast and central-west regions of Brazil.

2.4.2 *Genetic resources*

According to Freire Filho *et al.* (2011b), the genetic resources available for genetic improvement, the development of new populations and strains, as well as cowpea cultivars, are distributed in basic collections at international level, national level, active collections and working collections.

International base collections are generally located in international research institutes, which aim to guarantee the security of genetic material for the current and future demands of the breeding programs of the countries concerned. The *International Institute of Tropical Agriculture* (IITA) is responsible for housing the international cowpea base collection. Its germplasm bank contains the largest and most diverse collection in the world, with 15,122 unique specimens from 88 countries, representing 70% of African cultivars and almost half of global diversity (IITA, 2016).

The national cowpea base collection, on the other hand, is located at the National Center for Genetic Resources and Biotechnology (CENARGEN), in Brasilia (DF). This type of collection is usually located in a strategic location and belongs to a national research institution (FREIRE FILHO *et al.*, 2011b). CENARGEN has around 4,000 accessions and aims to meet the demands of genetic improvement programs in Brazil (WETZEL *et al.*, 2005 *apud* FREIRE FILHO *et al.*, 2011b).

By active collection, we mean one that aims to meet the demand for germplasm from the institution to which it belongs, as is the case with Embrapa Meio-Norte, located in Teresina (PI), which has around 3,500 accessions, and the Center for Agricultural Sciences at the Federal University of Ceara, in Fortaleza (CE), which has approximately 1,000 accessions. Both meet the demands of the institutions that make up the National Agricultural Research System (SNPA) (Table 3) (FREIRE FILHO *et al.*, 2011b).

Table 3 - Component institutions of the National Agricultural Research System.

Institution	Age/State
Embrapa Meio-Norte	Teresina, Piaui
Embrapa Rondbnia	Porto Velho, Rondonia
Embrapa Amazonia Occidental	Manaus, Amazonas
Embrapa Roraima	Boa Vista, Roraima
Embrapa Amapà	Macapa, Amapà
Embrapa Amazonia Orientai	Belém, Parǎ
Embrapa Cerrados ᵘ	Brasilia, Distrito Federai
Embrapa Tabuleiros Costeiros	Aracaju, Sergipe
Embrapa Agropecuaria Oeste	Dourados, Mato Grosso do Sul
Embrapa Agrobiologia	Seropedica, Rio de Janeiro
Embrapa Food Technology	Guaratiba, Rio de Janeiro
Embrapa Rice and Beans	Santo Antônio de Goiǎs, Goiàs
Embrapa Genetic Resources and Biotechnology	Brasilia, Distrito Federai
Embrapa Technology Transfer	Brasilia, Distrito Federai
Agricultural Research Company of Rio Grande do Norte - EMPARN	Natal, Rio Grande do Norte
Institute Agronòmico de Pernambura - IPA	Recife, Fernambuco
Bahia Agricultural Development Company - EBDA	Salvador, Bahia
Technological Education Center of the State of Ceará - CENTEC	Fortaleza, Ceará

⁽¹⁾Acao de pesqutsa realzada en∩ Gurupi, Tocantins

Finally, a working collection is one that has "restricted genetic variability and constitutes the genetic material handled day-to-day by the plant breeder" (FREIRE FILHO *et al.*, 2011b). In Brazil, Embrapa Meio-Norte, Embrapa Semiarido and the Instituto Agronòmico de Pernambuco have working collections of cowpea.

As for the methods of genetic improvement for cowpeas, the most widely used are: germplasm introduction, mass selection in local cultivars, individual plant selection with progeny testing in local cultivars, genealogical method, *single* seed descent method, *single pod descent* method and backcrossing method (FREIRE FILHO *et al.*, 2011b).

Table 4 - Breeding methods used and number of cowpea cultivars released commercially between 1903 and 2010.

Improvement method	Breeding				Total
	1ª Phase	2ª Phase	3ª Phase	4 ª Phase	
	(1903 a 1963)	(1963 a 1973)	(1973 a 1991)	(1991 a 2010)	
Introduction of gemnoplasm	Nd[1]		10	2	12
Mass selection		2	5	1	8
Individual plant selection with test of progeny		1		2	3
Genealogical			23	8	31
Offspring of a single seed				2	2
Single pod descent				15	15
Total		3	38	30	71

2.4.3 *Objectives and perspectives*

Broadly speaking, the short-term objectives of genetic improvement of the cowpea in Brazil are to improve its size (for both manual and mechanized harvesting), as well as its visual and culinary aspect, to increase productivity, stability, adaptability of the crop, protein and nutritional content of the grains, to develop cultivars adapted to all regions of Brazil and to increase resistance to pests, diseases, high temperatures and water stress (FREIRE FILHO *et al.*, 2011b).

Breeding work has enabled breeders to develop elite cultivars and improved strains, mainly in relation to increased productivity, resistance to pests and diseases, nematodes, invasive plants and tolerance to water and salt stresses (EHLERS & HALL, 1997 *apud* COSTA, 2010).

According to Bevilaqua *et al.* (2007), the development of research that evaluates the potential of creole and traditional varieties is extremely important. Just like the common bean (*Phaseolus vulgaris*), the caupi bean is mostly grown for subsistence farming, which, according to Embrapa Arroz e Feijao (2010), is mainly characterized by not buying seeds periodically, based on the use of the kernels obtained as seeds for several years. Despite their low productivity compared to genetically modified seeds, creole seeds are an excellent source of germplasm, which is often passed down from one generation of farmers to the next (EMBRAPA ARROZ E FEIJÂO, 2010).

It is important to note that the successive cultivation of the same germplasm increases the chance of mutations occurring and that organisms with adaptive advantages over others are preserved. Based on this characteristic, farmers, especially those who are more experienced with the crop grown, such as cowpeas, select the varieties that can possibly provide them with productive advantages (EMBRAPA ARROZ E FEIJÂO, 2010).

It is very common for creole or local varieties to have a combination of several genotypes, which is very relevant in the field of genetic improvement, since these organisms can contain sources of tolerance to various biotic and abiotic stresses, such as water stress, and are endowed with desirable agronomic characteristics (EMBRAPA ARROZ E FEIJÂO, 2010).

Since it was first introduced in Brazil in 2010, 71 genetically improved cowpea cultivars have been launched (Table 4), which is a very small number compared to cultivars of other annual crops grown in the country.

Despite the progress made in the cultivation of this legume in Brazilian regions, there is still a need for greater investment in research and breeding work to improve cowpea plants, especially with regard to its creole varieties.

2.5 Water stress

2.5.1 Concetto and importing, in cowpea cultivation

Among the abiotic stresses to which plants can be subjected, such as waterlogging, nutritional deficiency and salinity, drought, or water deficit, is one of the most relevant factors in terms of the loss of potential crop productivity, and can have drastic effects on plant growth and development.

The cowpea is a widely grown crop in the Brazilian semi-arid region, among other things because of its ability to tolerate water stress, when compared to other economically important legumes such as common beans and chickpeas (EMBRAPA SEMI-ARIDO, 2007). Because of this, i.e. because it is a drought-tolerant plant, it can be grown in a variety of climatic and soil conditions (EMBRAPA, 2011).

Its water demand varies from 300 to 450 mm/cycle, which depends on the local growing conditions, the cultivar used and the type of soil. It rarely exceeds 3.00 mm/day when the plant is in the early stages of development (EMPRAPA MEIO-NORTE, 2002).

Bezerra *et al*. (2003) state that water deficit affects practically all aspects of plant growth, including changes in the anatomy, physiology, morphology and biochemistry of the plant, developing in a very complex way. According to Stamford et al. (1990 *apud* EMBRAPA MEIO-NORTE, 2002), water stress even reduces the weight of the crop's root nodules and the nitrogen it accumulates, as well as the dry matter production of its aerial part, especially when the water deficit is imposed in the second and fifth weeks after sowing.

Although it is considered a drought-tolerant crop, it has been observed that the occurrence of water deficit, especially in the flowering and grain filling stages, can cause severe reductions in the productivity of cowpea grains (CORDEIRO *et al*. 1998; SANTOS *et al*. 1998 *apud* EMBRAPA MEIO-NORTE, 2002). Labanauskas *et al*. (1981 *apud* BEZERRA *et al*., 2003), analyzing the effect of water deficiency on the different stages of growth and grain production in the cowpea crop, found that low water availability during the flowering and pod formation phases reduced production by 44 and 29%, respectively, compared to plants that did not suffer a reduction in water availability. Shouse *et al.* (1981), studying the effect of water deficit on cowpea production, obtained higher yield reduction results. They found that water stress during the flowering and pod filling stages caused a reduction of between 35 and 69% in grain yield. Like Labanauskas *et al.* (1981), they found that during the vegetative stage, water stress had less influence on grain yield. Shouse *et al.* (1981); Ferreira *et al.* (1991); Fiegenbaum *et al.* (1991); Brito (1993 *apud* BEZERRA *et al.*, 2003); Guimaraes *et al.* (1996); Andrade *et al.* (1999 *apud* BEZERRA *et al.* 2003), working with beans, also observed that the flowering and grain filling stages are the most critical to water deficit.

Bezerra *et al.* (2003) found that water deficit statistically affected not only grain yield, but also the number of pods per plant and the number of grains per pod in the cowpea crop. However, pod length and the mass of 100 grains were not statistically affected by water deficit in the same study.

Bastos *et al.* (2002 *apud* BEZERRA *et al.*, 2003) evaluated the growth and development of caupi beans under different irrigation levels and observed a significant reduction in leaf area as the water stress intensified.

2.5.2 *Main cells and adaptive mechanisms of water stress in plants*

Despite the fact that drought is one of the most important factors in reducing crop productivity, plants have acquired a number of adaptive mechanisms over time, expressing their resistance in various ways. Three main strategies can be highlighted, as conceptualized by Turner *et al.* (2000): drought escape, dehydration tolerance and delayed dehydration. However, cultivated plants use more than one mechanism at a time to cope with water stress (AGBICODO *et al.*, 2009).

Lawan (1983 *apud* AGBICODO *et al.*, 2009) and Boyer (1996) state that both the mechanism of stomatal closure to reduce water loss through transpiration and the cessation of growth (drought escape mechanisms) and osmotic adjustment and slow, continuous growth (drought tolerance mechanisms) are drought adaptation mechanisms observed in cowpeas.

2.5.2.1 *Escape the drought*

This strategy consists of shortening the life cycle of the plant, which develops (germinates, grows, flowers and fruits) during periods of water availability in a shorter time than compared to its normal life cycle.

During drought escape, plants optimize the use of the limiting resources available during periods of water supply in order to complete their development before drought begins, thus escaping the drought period, which can occur after the rains.

According to Amudha & Balasubramani (2011), drought escape can only occur in annual plants and does not occur in perennial crops, which are often exposed to drought conditions.

2.5.2.2 *Tolerance to dehydration*

Drought tolerance can be defined as the ability of some plants to live, grow and produce satisfactorily with limited water availability in the soil or under periodic water deficiencies (ASHLEY, 1993 *apud* AGBICODO *et al.*, 2009). Nguyen *et al.* (1997) also characterized dehydration tolerance. For them, this occurs when the plant continuously maintains its physiological processes under conditions of water deficit.

Plant species endowed with this adaptive mechanism are able to carry out some artifices that allow them to survive. Engelbrecht & Kursar (2003) and Tyree *et al.* (2003) reported that they are able to reduce the extent of xylem vessel cavitation, as well as the rate of macromolecule oxidation, without altering leaf gas exchange variables. Wilson *et al.* (1980), in turn, observed that they are able to develop more rigid cell walls. Morgan (1984) stated that dehydration-tolerant C4 plants are able to increase the production of buliform cells and maintain the turgidity of their cells, even in the presence of low water potentials, through the accumulation of solu- tions.

Several factors and mechanisms act independently or together to enable plants to cope with water stress. Therefore, drought tolerance manifests itself as a complex trait (KRISHNAMURTHY *etal.*, 1996).

2.5.2.3 *Delayed dehydration*

According to Chaves *et al.* (2003), delaying dehydration can be achieved by maximizing access to water and minimizing its loss to the atmosphere. Pathan *et al.* (2004) and Uga *et al.* (2013) cited high biomass investment in the root system, increased root growth angle, production of very fine roots, high specific root length, small leaf area and efficient stomatal control of transpiration as characteristics associated with this adaptation strategy.

It is worth noting that the plant's choice of reproductive strategy depends on the plant's ability to adjust its phenotype appropriately and at a certain speed.

2.5.2.4 *Other adaptive mechanisms*

In addition to the adaptive mechanisms to water stress mentioned above, we can also mention the reduction in photosynthetic activity - which is due to the synchronized occurrence of various processes, such as stomatal closure and a reduction in the activity of photosynthetic enzymes - the accumulation of organic acids and changes in carbohydrate metabolism. Photosynthesis and cell growth are among the first processes to be affected by drought (CHAVES, 1991).

Condon & Hall (1997) observed that there was a relationship between the productivity of peanut, cowpea and common bean crops and the carbon isotope. Hall et al. (1997) and Condon & Hall (1997) positively associated genotypic differences in the productivity potential of cowpea grains with the carbon isotope, concluding that more productive genotypes have a higher photosynthetic rate, resulting in a higher internal concentration of carbon dioxide in their leaves, thus maintaining an association between the carbon isotope and crop performance.

Cruz de Carvalho *et al.* (1998) compared the physiological responses of cowpea and common bean genotypes in terms of stomata opening under water stress. They found that the cowpea genotypes kept their stomata partially open under this condition, with a smaller decrease in their photosynthetic rates than the common bean.

Several other mechanisms can partially explain how the cowpea crop is able to avoid extreme dehydration. During the vegetative stage, these mechanisms may be related to the low decrease in leaf water potential, even under extreme conditions of water deficit.

The lowest leaf water potential for the cowpea crop was -18 bar (-1.8 MPa) (TURK & HAL 1980; HALL & SCHULZE 1980 *apud* AGBICODO *et al.*), while for the peanut crop, which has high leaf water potentials when exposed to drought, it was -82 bar (-8.2 MPa) (TURNER *et al.*, 2000).

Cowpeas are also able to change the position of their leaflets when subjected to water stress, which is a drought prevention mechanism. They become paraeliotropic, being able to orient themselves parallel to the sun's rays, when subjected to dry soil, making it cooler and thus transpiring less (SHACKEL & HALL, 1979 *apud* AGBICODO *et al.* 2009), which helps to minimize water loss and maintain water potential.

2.5.2.4.1 Cellular level

Water deficit is capable of promoting the formation of reactive oxygen species (ROS) and reactive nitrogen species (RNS), such as hydrogen peroxide (H_2O_2) and nitric oxide (NO), respectively (FERRARI, PAZ & SILVA, 2015). At high levels, these substances can cause oxidative damage. Such damage can be more evident in the leaves, in the case of damage caused by ROS (such as photoinhibition), or more evident in the roots, in the case of damage caused by RNS.

2.5.2.4.2 At the molecular level

One strategy used by plants at a molecular level is the induction of gene expression, in this case of genes capable of responding to water stress. Scholars claim that when these genes are expressed, products such as functional metabolites and proteins are encoded in response to stressful conditions, thus conferring tolerance to the stress suffered by the plant.

In conditions of water stress, for example, the genes that are induced protect the cells against dehydration, through the production of important metabolic proteins, as well as through the genetic regulation that carries out signal transduction in response to water deficit. Thus, the introduction of genes capable of being induced by stress, by means of gene transfer, has increased the tolerance of plants to water stress (ZHANG *et al.*, 2004; UMEZAWA *et al.*, 2006).

A series of genes capable of being induced by stress have recently been identified and their products classified. Subsequently, these genes were cloned and characterized in various plant species (NEPOMUCENO *et al.*, 2001). Umezawa *et al.* (2006) used proteins capable of being induced by stress, whose functions were known (such as aquaporins), to make changes in plants. However, the development of drought-tolerant organisms has not been an easy task, since drought is a condition controlled by several genes.

Therefore, there is still a need for more studies and research in this area, with the aim of better understanding the genes involved in this process, as well as the products generated by their expression, so that it becomes possible to better characterize the mechanisms that promote the adaptation of organisms to conditions of water deficit (BRAY, 2004).

Two important gene families responsive to water stress in plants are DREB (*Dehydration Responsive Element Binding*) and ERF (*EthyleneResponsive Factor*), which are part of the AP2/EREBP (*Ethylene- Responsive-Element-Binding Protein*) superfamily of transcription factors exclusive to plants. For more than ten years, members of these families have been highlighted for their role in water stress tolerance, through ABA-dependent and -independent metabolic pathways, as well as for their stress-responsive regulation, which encompasses hundreds of target genes (BENKO-ISEPPON *et al.*, 2011). However, studies carried out with transgenic plants have shown that their expression requires close tuning, since the constitutive "overexpression" of the DREB/CBF pathway led to drastic defects in the development of the transformants, although it was accompanied by greater tolerance to cold, salt and drought (KASUGA *et al.* 1999; SINGH *et al.*, 2002; SHINOZAKI, YAMAGUCHI-SHINOZAKI & SEKI, 2003).

2.5.3 Evaluation methods for cowpea plants subjected to water stress

Agbicodo *et al.* (2009) classified six methods as the most suitable for evaluating a large number of cowpea strains for drought tolerance. These are: determining chlorophyll fluorescence, measuring stomatal conductance, measuring abscisic acid, measuring free proline levels, delaying leaf senescence and screening in a wooden box for drought tolerance at the seedling stage, which was used in this work.

2.5.3.1 Wooden box method for evaluating seedlings for drought tolerance

Singh *et al.* (1999) suggested that the different organs of the cowpea plant (root, stem and leaf) should be used to assess drought tolerance. According to them, different tissues show different responses to abiotic stress and should therefore be studied separately. This could allow the identification of genetic factors in specific tissues when subjected to drought and the elucidation of parts of the drought response pathways, thus facilitating genetic improvement (AGBICODO *et al.*, 2009).

A simple method used to identify drought tolerance in cowpea seedlings is the "wooden box technique" (FIGURE 3).

Figure 3 - Boxes used to screen cowpea varieties in response to water stress.

Source: AGBICODO *et al.* (2009).

This method, which has proven to be efficient in screening for drought tolerance in different crop species, is able to eliminate the influences of the root system on drought tolerance and allows non-destructive visual identification of the effects of dehydration (SINGH *et al.*, 1999; TOMAR & KUMAR 2004; SLABBERT *et al.* 2004; EWANSIHA & SINGH, 2006). In addition, the results obtained with this method have shown a close similarity with the results obtained from experiments carried out in the field for various crops, both at the seedling stage and at the reproductive stage (AGBICODO, 2009). One of the main objectives of this method is to keep the plants' root systems separate and eliminate competition for water between the genotypes, while maintaining low space requirements (MUCHERO *et al.*, 2008).

The wooden box method has been used to evaluate cowpea varieties that are sensitive and tolerant to severe water stress, and its use has generated contrasting results (AGBICODO *et al.*, 2009), i.e. it has revealed differences in the behavior of varieties in terms of drought tolerance.

Singh, Kodomi & Terao (1999), working with various genotypes of cowpea, used the wooden box sorting method. The researchers carried out the experiment in a screened environment and the boxes were elevated on trellises to avoid contact with rainwater and soil, respectively. The boxes (130 cm long, 65 cm wide and 15 cm deep) used by the researchers were lined with polyethylene sheets and filled with a substrate made up of a mixture of soil and sand in a 1:1 ratio. The substrate was filled to a depth of 12 cm, leaving about 3 cm of space at the top for watering.

Polyethylene sheeting lining the entire inside of the boxes (sides and bottom) ensured even water distribution. A bubble level was used to ensure a flat surface on the ground, after which the boxes were irrigated.

Equidistant holes were drilled in straight lines, approximately 10 cm apart and 5 cm between plants in the same line. Six cowpea seeds were sown per hole and, after germination, they were thinned out, keeping only one plant per hole. Each box contained one row of each of the cowpea varieties, with 12 plants (FIGURE 4). The boxes were watered daily, using a fixed volume of water, until the first pair of trifoliate leaves emerged. After that, watering was suspended.

Figure 4 - Wooden boxes used by Singh, Kodomi & Terao.

Source: Adapted by the author from Singh, Kodomi & Terao (1999).

The researchers then counted the number of plants with a permanent gait each day for each variety, until all the plants in the rows had died. Then, watering was resumed in order to determine the regeneration percentages for each cowpea variety.

Based on the days it took for the plants to wilt and the percentage of recovery, the researchers classified the cowpea varieties as drought tolerant or susceptible.

Despite having many positive points, this methodology has some drawbacks, which could be observed during the experiment. These include the need for daily data collection, which should preferably be carried out by the same evaluator, in order to avoid discrepancies in the results observed, since the method can be considered qualitative. In addition, the occurrence of pests, diseases and death affecting certain plants in the box can consequently have positive or negative effects on the development and growth of the other plants in the box, which can influence the results obtained by the observer.

3. MATERIAL AND METHODS

3.1 Location and conduct of the experiment

The work was carried out in a greenhouse (FIGURE 5) belonging to the Federal University of Ceara (UFC), in Fortaleza (CE), from October 22 to November 30, 2015, lasting 39 days (32 days with irrigation suspended). 30 varieties of cowpea were used.

Figure 5 - General view of the experimental area.

Source: Author (2015).

The seeds were sown in polyethylene boxes, raised at a distance of 9 cm from the ground, with dimensions of 53 cm long, 37 cm wide and 24 cm high. These were filled with a homogeneous and sterile mixture in a ratio of 6:3:1 of sand, earthworm humus and vermiculite, respectively. In each box, 8 varieties were sown with 6 plants, and each variety was repeated 4 times per trial. The spacing used between rows and between plants was approximately 6.0 cm and 7.0 cm, respectively (FIGURE 6).

Before sowing, the seeds were sterilized for 1 minute with sodium hypochlorite (NaClO) diluted to 0.5%. They were then distributed on two sheets of germ paper moistened with distilled water, in an amount equivalent to 2.5 times the mass of the dry paper, and transferred to a BOD (*Biochemical Oxygen Demand*) germinator at a temperature of 25 °C. After two days, when they had germinated, they were transferred to polyethylene boxes. The boxes were irrigated daily with 1.5 liters of distilled water.

Figure 6 - Schematic of the plots in each box.

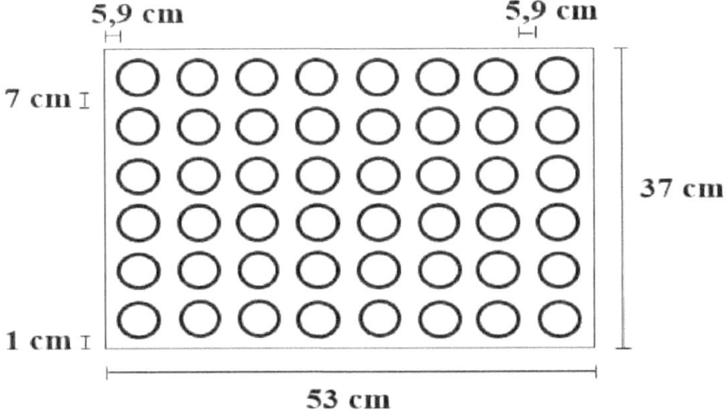

Source: Prepared by the author (2016).

3.2 Evaluated plant material

The varieties evaluated in this work are local varieties of cowpea collected in different municipalities in the state of Ceara, preferably grown by family farmers. Twenty-five local varieties were evaluated, in addition to CE-031, CE-315 and CE-930, which come from the UFC's Active Germplasm Bank (BAG) of cowpea. The Pingo de Ouro 1,2 variety was used as a control variety for tolerance to water stress, while the Santo Inacio variety was used as a control variety for susceptibility to water stress (TABLE 5).

Table 5 - Varieties used.

N°	Common name of the variety	Place of collection (Origin)
1	CE-031 (Pitiuba)	-
2	CE-315 (Tvu 2331)	-
3	CE-930 (Pingo de Ouro)	-
4	Pingo de Ouro 1,2	-
5	Saint Ignatius	-
6	Chumbinho	Barbalha/CE
7	Maranhao	Barbalha/CE
8	Willer's Black Beans	Unknown
9	Canapu	Deputy Irapuan Pinheiro/CE

10	Drop of Gold	Deputy Irapuan Pinheiro/CE
11	Snap beans	Deputy Irapuan Pinheiro/CE
12	Always green	Guaraciaba do Norte/CE
13	Bush beans - red	Guaraciaba do Norte/CE
14	Always green	Guaraciaba do Norte/CE
15	String beans	Guaraciaba do Norte/CE
16	Pitiuba	Morada Nova/CE
17	Drop of gold	Morada Nova/CE
18	Epace 10	Morada Nova/CE
19	Beans from Bahia	Parambu/CE
20	Cojò	Parambu/CE
21	Saint Ignatius	Parambu/CE
22	Zé Artur	Paramoti, Monte Pedal/CE
23	Meadow gnat	Paramoti, Monte Pedal/CE
24	Black face	Sao Benedito/CE
25	Xique-xique	Sao Benedito/CE
26	Cheat on a woman	Farias Brito, Cariri/CE
27	String beans	Farias Brito, Cariri/CE
28	Owl's eye bean	Farias Brito/CE
29	Evergreen beans	Farias Brito/CE
30	Canapu - Light	Farias Brito/CE

Source: Prepared by the author (2016).

3.3 Experimental design

The experimental design used was randomized interaction design (CID) with 30 treatments and four replications. The plots consisted of six plants and were arranged in rows within each box (FIGURE 7).

Figure 7 - Layout of plants per plot.

Source: Author (2015).

To assess the effects of water stress, the boxes were irrigated until the seventh day after sowing, when the first pair of true leaves appeared, after which the plants had their irrigation suspended. The data collected corresponded to the number of plants with severe wilt, which was recorded daily.

At the end of the data collection, statistical analysis was carried out and the Skott-Knott test was applied at a 5% probability level, using Assistat - Statistical Assistance - *software*, with the aim of identifying and grouping the drought-susceptible and drought-tolerant varieties, and the Dunnett test, using Action Stat 3.1 *software*, in order to compare the means of the local varieties with the means of the control varieties.

4 RESULTS AND DISCUSSION

4.1 Results

4.1.1 Scott-Knott test

There were no statistically significant differences between the treatments in terms of drought tolerance (TABLE 1), i.e. when compared to each other using the Skott-Knott test at a 5% probability level, the means of the treatments did not differ statistically from each other (GRAPH 1).

Table 1 - *Analysis table.*

FV	GL	SQ	QM	F
Treatments	29	415.18368	14.31668	1.0928 ns
Waste	90	1179.06634	13.10074	
Total	119	1594.25002		

ns: not significant ($p >= 0.05$)

Source: Adapted from Assistat software by the author (2016).

Graph 1 - Average number of days to severe wilting.

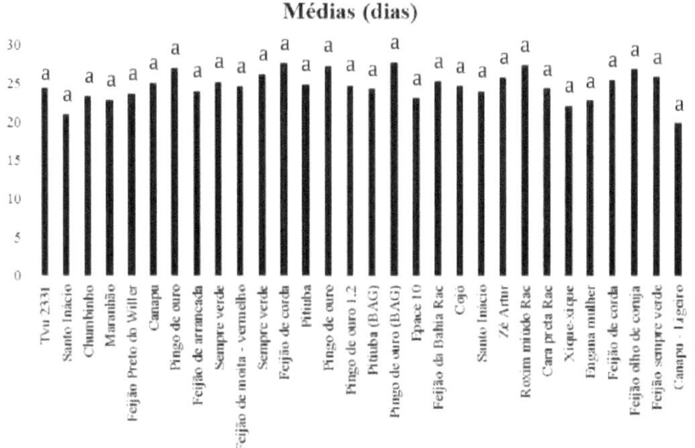

Averages followed by the same letter are not statistically different. Source: Prepared by the author (2016).

4.1.2 Dunnett test

4.1.2.1 Comparison between the averages of the Iocais varieties and the Santo Indcio variety

The Santo Inacio variety did not show the expected drought susceptibility behavior, and did not differ significantly from the others at the 5% probability level using Dunnett's test (GRAPH 2). There would be a significant difference between the means of the treatments if the confidence intervals of the local varieties (represented in blue) did not touch the dashed red line.

Graph 2 - Confidence intervals between local varieties and the Santo Inacio variety.

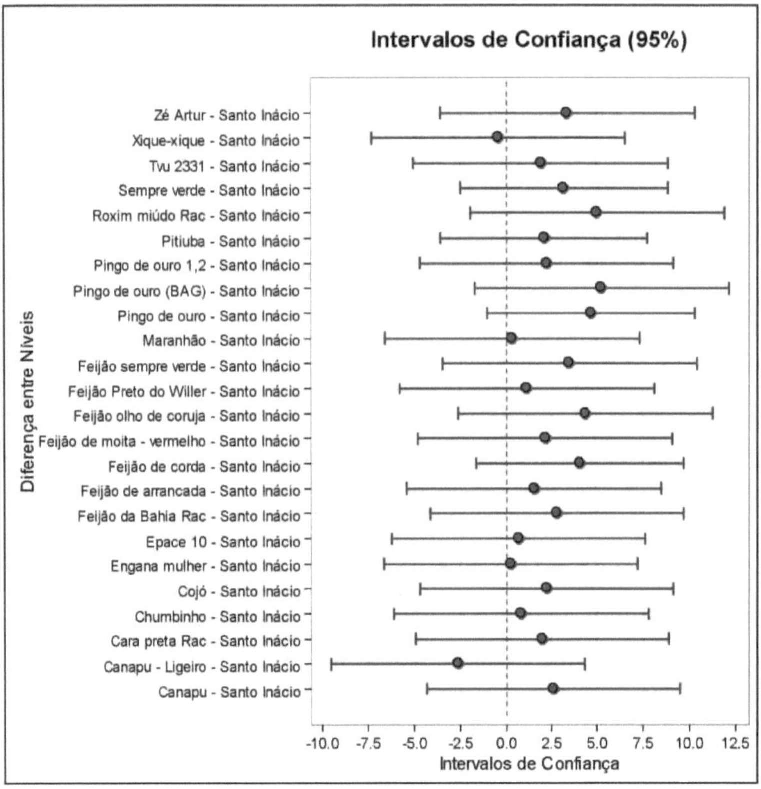

Source: Action Stat software (2016).

4.1.2.2 Comparison between the averages of the local varieties and the variety

Drop of gold 1, 2

The Pingo de Ouro 1,2 variety, used as a control variety for drought tolerance, did not differ significantly from the other treatments at the 5% probability level using Dunnett's test (GRAPH 3). There would have been a significant difference between the means of the treatments if the confidence intervals of the local varieties (represented in blue) did not touch the dashed red line.

Graph 3 - Confidence intervals between local varieties and the Pingo de ouro 1,2 variety.

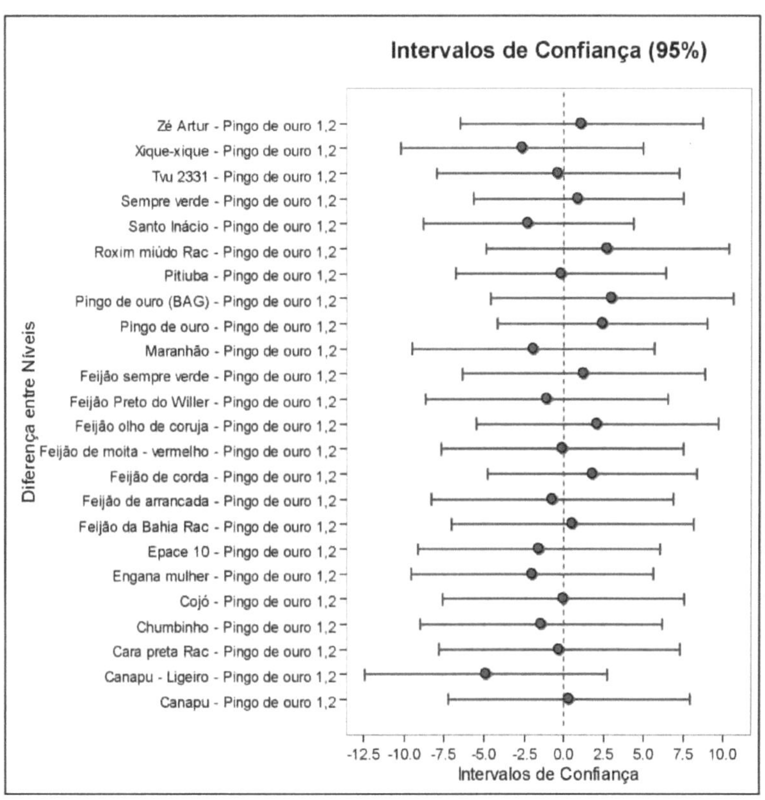

Source: Action Stat software (2016).

4.2 Discussion

The results obtained can be justified by the fact that only one comparative parameter was used in this experiment, in this case the variable number of days to severe wilt in cowpea plants.

According to Nascimento *et al.* (2011), "there is no single physiological variable that, on its own, is indicative of drought tolerance", which can be complemented by Nogueira *et al.* (2001), who state that the ideal is for more than one variable to be assessed, in particular important variables in assessing the responses of plant species

31

to water stress, such as stomatal conductance, water potential, temperature and leaf transpiration.

Mendes *et al.* (2007) state that, when assessing the water status of plants subjected to water deficiency, both in the vegetative and reproductive phases, cowpea cultivars have been shown to suffer significant reductions in leaf water potential, stomatal conductance and leaf transpiration, generating a consequent increase in leaf temperature.

Thus, in order to obtain different responses regarding susceptibility and tolerance to drought, with a possible grouping of varieties into susceptible or tolerant, it would be advisable to carry out the experiment again, adding other evaluation parameters. To this end, the following analyses could be carried out: stomatal conductance, temperature and leaf transpiration, using the IRGA (*Infra-red Gas Analyzer*) apparatus; and, instead of evaluating water potential, evaluating relative water content and dry mass, as these are more feasible and non-destructive assessments.

5 CONCLUSION

This study did not find any varieties that were tolerant or sensitive to water stress among the Iocal cowpea varieties evaluated using the "number of days to permanent wilting" criterion.

Therefore, it is important to carry out studies in which the same cowpea varieties are evaluated for more variables that confirm their tolerance or susceptibility to water stress, in order to complement the results obtained in this study.

REFERENCES

AGBICODO, E. M.; C. A. FATOKUN; S. MURANAKA, R. G. F. VISSER; C. G. LINDEN VAN DER. Breeding drought tolerant cowpea: constraints, accomplishments, and future prospects. **Euphytica**, v. 167, p. 353-370, 2009.

AGEITEC - EMBRAPA TECHNOLOGICAL INFORMATION AGENCY. **The economic importance of cowpeas.** Available at: <http://www.agencia.cnpti-a.embrapa.br/gestor/feijao-caupi/arvore/CONTAG01_14_510200683536.html>. Teresina, 2016. Accessed on: May 21, 2016.

ALMEIDA, Jorge. **Evaluation of cowpea cultivars used in the Seed Distribution Program**. Cruz das Almas, 2014. Available at:

<http://www.seagri.ba.gov.br/sites/default/files/5_pesquisa_agricola02v9n3_0.pdf>. Accessed on: 02 Apr. 2016.

ALVES, J. M. A.; ARAÙJO, N. P.; UCHÔA, S. C. P.; ALBUQUERQUE, J. A. A.; SILVA, A. J.; RODRIGUES, G. S.; SILVA, D. C. O. Avaliaçâo agroeconômica da produçâo de cultivares de feijâo-caupi em consórcio com cultivares de mandioca em Roraima. **Agro@mbiente On-line**, Boa Vista, v. 3,n. 1, p. 15-30, Jan./Jun. 2009.

AMUDHA, J.; BALASUBRAMANI, G. Recent molecular advances to combat abiotic stress tolerance in crop plants. **Biotechnology and Molecular Biology Review**, v. 6, n. 2, p. 31-58, 2011.

BASTOS, E. A.; NASCIMENTO, S. P. do; SILVA, E. M. da; FREIRE FILHO, F. R.; GOMIDE, R.L. Identification of cowpea genotypes for drought tolerance. **Revista Ciência Agronòmica**,v. 42,n. 1,p. 100-107. 2011.

BENKO-ISEPPON, A. M.; SOARES-CAVALCANTI, N. M.; BERLARMINO, L. C.;

BEZERRANETO, J. P.; AMORIM, L.L.B. *etal*. Prospecting for Genes of Drought and Salinity Resistance in Native and Cultivated Plants. **Magazine**

Brasileira de Geografia Fisica, v. 6, p. 1112-1134, 2011.

BEVILAQUA, G. A. P.; SILVA, S. D. A.; ANTUNES, I. F.; BARBIERI, R. L.; GALHO, A. M.; BAMMANN, I. Seed bank of creole and traditional varieties of temperate climate family farming. **Revista Brasileira de Agroecologia**, v. 2,n. 1, p. 654-657, feb. 2007.

BEZERRA, F. M. L.; ARARIPE,. A. E.; TEÔFILO, E. M.; CORDEIRO, L. G.; SANTOS, J. J. A. dos. Cowpea and water deficit in its phenological phases.

Revista Ciência Agronòmica, v. 34, n. 1, p. 5-10, 2003.

BOYER, J. S. Advances in drought tolerance in plants. **Advances in Agronomy**, v. 56,p. 189-218, 1996.

BRAY, E. A. Genes commonly regulated by water-deficit stress in Arabidopsis thaliana. **Journal ofExperimental Botany**, v. 55, p. 2331-2341, 2004.

CHAVES, M.M. Effects of water deficits on carbon assimilation. **Journal of Experimental Botany**, v. 42, p. 1-16, 1991.

CHAVES, M. M.; MAROCO, J.P.; PEREIRA, J.S. Understanding plant responses to drought- from genes to the whole plant. **Functional Plant Biology**, v. 30, p. 239-264, 2003.

COELHO, D. S.; MARQUES, M. A. D.; SILVA, J. A. B.; GARRIDO, M. S.; CARVALHO, P. G.S. Physiological responses in cowpea varieties submitted to different shading levels. **Revista Brasileira de Biociências**, Porto Alegre,v. 12,n. 1,p. 14-19,jan./mar. 2014.

CONAB - NATIONAL SUPPLY COMPANY. **Monitoring the Brazilian grain harvest**. Brasilia, v. 3, n. 8 p. 1-178, May 2016. Available at:< http://www.conab.gov.br/OlalaCMS/uploads/arquivos/16_05_19_ll_58_17_boleti m_grains_may_ 2016_-_final.pdf>. Accessed on: May 21, 2016.

CONDON, AG; HALL, A.E. Adaptation to diverse environments: variation in wateruse efficiency within crop species. In: JACKSON, L.E. (ed) Ecology in agriculture. **Academic Press**, San Diego, p. 79-116, 1997.

COSTA, Eva Maria Rodrigues. **Genetic divergence between African cowpea lines (Vigna Unguiculata (L.) Walp.) through morpho-agronomic and molecular characterization.** 2010. 100f. Dissertation (Master's Degree in Agronomy) - Faculty of Agronomy, Rural University of Pernambuco, Recife, 2010.

CRUZ DE CARVALHO, M. H.; LAFFRAY, D.; LOUGUET, P. Comparison of the physiological responses of Phaseolus vulgaris and Vigna unguiculata cultivars when submitted to drought conditions. **Environmental and Experimental Botany**, v. 40, p. 197-207, 1998.

EMBRAPA-Brazilian Agricultural Research Corporation. **Botanical, morphological and agronomic characterization of caupi cultivars collected in the state of Acre**. Rio Branco, 1997. Available at:< http://ainfo.cnptia.embrapa.br/digital/bitstream/CPAF-AC/1212/1/bp17.pdf>. Accessed on: May 19, 2016.

EMBRAPA - Brazilian Agricultural Research Corporation. Production Systems - **Cultivation of cowpea (Vigna unguiculata (L.) Walp)**. Teresina, 2002

EMBRAPA - Brazilian Agricultural Research Corporation. Documents - **Active collection of cowpea germplasm (Vigna unguiculata (L.) Walp.) and other species of the genus Vigna, from Embrapa Meio-Norte, from 1976 to 2003**.

Teresina, 2011.

EMBRAPA RICE AND BEANS - Brazilian Agricultural Research Corporation. Documents - **Collection of traditional (creole) varieties of common bean (Phaseolus vulgaris L.) in the northern region of Rio Grande do Sul.** Santo Antonio de Goias, 2010.

EMBRAPA MEIO-NORTE - Brazilian Agricultural Research Corporation. Production

Systems - **Cultivation of cowpea (Vigna unguiculata (L.) Walp).** Teresina, 2002.

EMBRAPA SEMI-ARIDO - Brazilian Agricultural Research Corporation.

Documents - **Genetic Improvement of Cowpea at Embrapa SemiArido.** Petrolina, 2007.

ENGELBRECHT, B. M.; KURSAR, T. A. Comparative drought-resistance of seedlings of28 species of co-occurring tropical woody plants. **Oecologia**, v. 136, p. 383-393,2003.

EWANSIHA, S. U.; SINGH, B. B. Relative drought tolerance of important herbaceous legumes and cerealsin the moist and semi-arid regions of West Africa. **Journal of Food, Agriculture and Environment**, v. 4, p. 188-190, 2006.

FEHR, W. R.; FEHR, E. L.; JESSEN, H. J. **Principles of cultivar development: theory and technique.** New York: Macmillan, 1987. v. 1, p. 319-327.

FERRARI, E.; PAZ, A. da.; SILVA, A. C. da. Hydric deficit in soybean metabolism in early sowings in Mato Grosso. **Nativa**, v. 3, n. 1, p. 67-77,jan./mar. 2015.

FERREIRA, L. G. R.; COSTA, J. O.; ALBUQUERQUE, I.M.DE. Water stress in the vegetative and reproductive phases of two caupi cultivars. **Pesquisa Agropecua- ria Brasileira**, Brasilia, v. 26, n. 7, p. 1049-55, 1991.

FREIRE FILHO, F. R.; CARDOSO, M. J.; ARAÙJO, A. G. Caupi: scientific nomenclature and common names. **PesquisaAgropecuaria Brasileira**, Brasilia, DF, v. 18, n. 12,p. 1369-1372, dez. 1983.

FREIRE FILHO, F. R.; RIBEIRO, V. Q.; ROCHA, M. M.; SILVA, K. J. D.; NOGUEIRA, M. S. R.; RODRIGUES, E. V. **Cowpea in Brazil:** production, genetic improvement, advances and challenges. 84 p, 21 ed., Teresina, 2011b.

FREIRE FILHO, F. R.; RIBEIRO, V. Q.; ROCHA, M. M., SILVA, K. J. D.;

NOGUEIRA, M. S. R.; RODRIGUES, E. V. Production, genetic improvement and potential of cowpea in Brazil. **IV Biofortification Meeting**, Teresina, 2011a.

FREIRE FILHO, F. R.; RIBEIRO, V. Q.; SITTOLIN, I. M.; SILVA, S. M. S. e. Productivity of erect and semi-erect caupi lines in a cerrado environment. In: CONGRESSO BRASILEIRO DE MELHORAMENTO DE PLANTAS, 2001, Goiânia. **Proceedings...** SantoAntonio de Goias: Embrapa Rice and Beans, 2001.

FIEGENBAUM, V.; SANTOS, D. S. B. DOS; MELLO, V. D. C.; SANTOS FILHO, B. G. DOS; TILLMANN, M. A. A.; SILVA, J. B. Influence of water deficit on the yield components of three bean cultivars. **Pesquisa Agropecuaria Brasileira**, Brasilia, v. 26, n. 2, p. 275-80, 1991.

GUIMARAES, C. M.; STONE, L. F.; BRUNINI, O. Adaptation of the bean plant (Phaseo- Ius vulgaris L.) to drought. **Pesquisa Agropecuaria Brasileira**, Brasilia, v.31,n.7, p. 481-488, 1996.

HALL, A. E.; THIAW, S.; ISMAIL, A. M.; EHLERS, J.D. **Water-use efficiency and drought adaptation of cowpea**. In: SINGH, B.B. (ed) Advances in cowpea research. IITA, Ibadan, p. 87-98, 1997.

IITA-INTERNATIONAL INSTITUTE OF TROPICALAGRICULTURE. **Cowpea**. Ibadan, 2016. Available at: <http://www.iita.org/cowpea>. Accessed on: April 25, 2016.

KASUGA, M.; LIU, Q.; MIURA, S.; YAMAGUCHI-SHINOZAKI, K.; SHINOZAKI, K. Improving plant drought salt and freezing tolerance by gene transfer of a single stress-inducible transcription factor. **Nature Biotechnology**, v. 17, p. 287-291, 1999.

KRISHNAMURTHY, L. C.; JOHANSEN, C.; ITO, O. **Genotypic variation in root system development and its implication for drought resistance in Chickpea.** In: O. ITO; C. JOHANSEN; J. J. ADU-GYAMFI; K. KATAYAMA; J. V. D. K. KUMAR RAO; T. J. REGO (eds.), Dynamics ofRoots and Nitrogen in Cropping Systems of the Semi-Arid Tropics. Japan International Research Center for Agricultural Sciences, p. 235-250, 1996.

LABANAUSKAS, C.K.; SHOUSE, P.; STOLZY, L/E. Effects of water stress at various growth stages on seed yield on nutrient concentrations of fieldgrowncowpeas. **Soil Science**, Baltimore, v. 131,n.4,p. 249-256, 1981.

MENDES, R. M. S.; TAVORA, F. J. A. F.; PINHO, J. L. N.; PITOMBEIRA, J. B. Source-drain relationships in cowpea submitted to water deficiency. **Science Agronòmica**, v. 38, p. 95-103, 2007.

MORGAN, J. M. Osmoregulation and water stress in higher plants. **Annual Review of Plant Physiology and Plant Molecular Biology**, v. 35, p. 299-319, 1984.

MUCHERO, W.; EHLERS, J. D.; ROBERTS, P. A. Seedling stage drought-induced phenotypes and drought-responsive genes in diverse cowpea genotypes. **Crop Science**, v. 48,p. 541-552, 2008.

NASCIMENTO, S. P.; BASTOS, E. A.; ARAÙJO, E. C. E.; FREIRE FILHO, F. R.; SILVA, E. M. Water deficit tolerance in cowpea genotypes. **Revista Brasileira de EngenhariaAgricola e Ambiental**, Campina Grande, v. 15, n. 8, p. 853-860, 2011.

NEPOMUCENO, A.L.; NEUMAIER, N.; FARIAS, J.R. B.; OYA, T. Drought tolerance in plants. **Biotecnologia, Ciència e Desenvolvimento**, v. 23, p. 12-18, 2001.

NOGUEIRA, R. J. M. C.; MORAES, J. A. P V.; BURITY, H. A.; BEZERRANETO, E. Changes in leaf vapor diffusion resistance and water relations in acerola trees subjected to water deficit. **Revista Brasileira de Fisiologia Vegetal**, v. 13,p. 75-87,2001.

NGUYEN, T. T. T.; KLUEVA, N.; CHAMARECK, V.; AARTI, A.; MAGPANTAY, G.; MILLENA, A. C. M.; PATHAN, M. S.; NGUYEN, H. T. Saturation mapping of QTL regions and identification of putative candidate genes for drought tolerance in rice. **Molecular Genetics & Genomics**, v. 272, p. 35-46, 2004.

OLIVEIRA, A. P.; SOBRINHO, J. T.; NASCIMENTO, J. T.; ALVES, A. U.; ALBUQUERQUE, I. C.; BRUNO, G. B. Evaluation of cowpea lines and cultivars in

Areia, PB. **Horticultura brasileira**, Brasilia, v. 20, n. 2, p. 180-182, jun. 2002.

PATHAN, M.S.; SUBUDHI, P.K.; COURTOIS, B.; NGUYEN, H.T. **Molecular dissection of abiotic stress tolerance in sorghum and rice.** In Physiology and Biotechnology Integration for Plant Breeding. Edited by NGUYEN, H. T.; BLUM, A. Marcel Dekker, p. 525-569, 2004.

SEAB - SECRETARY OF STATE FOR AGRICULTURE AND Supply. **Beans**: Analysis of the Agricultural Situation. Paraná, 2015. Available at:

<http://www.agricultura.pr.gov.br/arquivos/File/deral/Prognosticos/2016/_feijao_201 5_16.pdf>. Accessed on: 09 Apr. 2016.

SILVA, P.S.L.; OLIVEIRA, C.N. Yields of green and mature beans from caupi cultivars. **Horticultura Brasileira**, Brasilia, v. 11, n. 2, p. 133-135, 1993.

SINGH, B. B.; MAI-KODOMI, Y.; TERAO, T. A simple screening method for drought tolerance in cowpea. **Indian Journal of Genetics and Plant Breeding**, v. 59,p.211-220, 1999.

SINGH, K. B.; FOLEY, R. C.; ONATE-SANCHEZ, L. Transcription factors in plant defense and stress response. **Current Opinion in Plant Biology**, v. 5, p. 430-436, 2002.

SHINOZAKI, K.; YAMAGUCHI-SHINOZAKI, K.; SEKI, M. Regulatory network of gene expression in the drought and cold stress responses. **Current Opinion in Plant Biology**, v. 6, p. 410-417, 2003.

SHOUSE, P.; DASBERG, S.; JURY, W. A.; STOLZY, L. H. Water deficit effects on water potential, yield, and water use cowpeas. **Agronomy Journal**, Madison, v.73,p. 333-336, 1981.

SLABBERT, R.; SPREETH, M.; KRUGER, G. H. J. Drought tolerance, traditional crops and biotechnology: breeding towards sustainable development. **South African Journal of Botany**, v.70,p. 116-123, 2004.

TOMAR, S.M. S.; KUMAR, G. T. Seedling survivability as a selection criterion for drought tolerance in wheat. Plant Breeding, v. 123,p. 392-394, 2004.

TURNER, N. C.; WRIGHT, G. C.; SIDDIQUE, K. H. M. Adaptation of grain legumes (pulses) to water limited environments. **Advances in Agronomy**, v.71: 193231,2000.

TYREE, M. T.; ENGELBRECHT, B. M.; VARGAS, G.; KURSAR, T. A. Desiccation tolerance of five tropical seedlings in panama. Relationship to a field assessment of drought performance. **Plant Physiology**, v. 132, p. 1439-1447, 2003.

UGA, Y.; SUGIMOTO, K.; OGAWA, S.; et al. Control of root system architecture by DEEPER ROOTING 1 increases rice yield under drought conditions. **Nature Genetics**, v. 45, p. 1097-1102, 2013.

UMEZAWA, T.; FUJITA, M.; FUJITA, Y.; YAMAGUCHI-SHINOZAKI, K.; SHINOZAKI, K. Engineering drought tolerance in plants: discovering and tailoring genes to unlock the future. **Current Opinion in Biotechnology**, v. 17, p.113-122, 2006.

UNIFEIJÂO. **Production history**. Sao Paulo, 2016. Available at: <http://unifeijao.com.br/site2013/feijao_brasil.php?txt=8> Accessed on: May 21, 2016.

WANDER, A. E. Brazilian production and participation in the international market for cowpea. **III National Cowpea Congress**. Recife, 2013.

WILSON, J. R.; LUDLOW, M. M.; FISHER, M. J.; SCHULZE, E. D. Adaptation to water stress of the leaf water relation of four tropical forage species. **Australian Journal of Plant Physiology**, v. 7, p. 207-220, 1980.

ZHANG, J. Z.; CREELMAN, R. A.; ZHU, J. K. From laboratory to field. Using information from Arabidopsis to engineer salt, cold, and drought tolerance in crops. **Plant Physiology**, v. 135,p. 615-621, 2004.

Printed by Books on Demand GmbH, Norderstedt / Germany